WORMS

CREEPY CRAWLERS

Lynn Stone

The Rourke Book Co., Inc.
Vero Beach, Florida 32964

PHOTO CREDITS
Cover © James P. Rowan; Title page and page 7 © James A.
Robinson; pages 4 and 8 © Michael Cardwell; pages 10, 15 and
18 © Breck P. Kent; page 12 © Frank Balthis; pages 13, 17 and 21
© Lynn M. Stone

Library of Congress Cataloging-in-Publication Data

Stone, Lynn M.
   Worms / by Lynn Stone.
      p. cm. — (Creepy crawlers)
   Includes index.
   ISBN 1-55916-160-4
   1. Worms—Juvenile literature.
I. Title II. Series: Stone, Lynn M.   Creepy Crawlers
QL 386.S77 1995
595.1—dc20
                                              95–16559
                                                  CIP
                                                      AC

Printed in the USA

TABLE OF CONTENTS

WORMS

Everyone knows the earthworm of lawns and gardens. To fishermen, the earthworm is an old friend.

Earthworms, though, are just one group of worms. Many other kinds of worms, such as flatworms, tapeworms, roundworms, and **marine** (muh REEN) worms live throughout the world. Most of these groups of worms are not related to the earthworms.

All worms are soft and long-bodied. Some are flat. Others are round. All worms are **invertebrates** (in VERT uh brayts), too—animals without bones.

A feather duster worm pokes out of its undersea tube

WHAT WORMS LOOK LIKE

Thousands of **species** (SPEE sheez), or kinds, of worms are shaped like pencils. Other living things with this shape have often been named after worms, such as worm lizard, worm snake, and worm vine.

Thousands of other worms, however, have flat, ribbonlike bodies. Worm shapes—round or flat— help them burrow, swim, or crawl.

Certain groups of worms are tiny. They can be seen only through a microscope. Other worms reach 30 feet in length!

This tapeworm parasite made its home in a cat until the cat threw it up

WHERE WORMS LIVE

Worms live in many different kinds of places. Earthworms may burrow nine feet into soil. Marine worms, the earthworm's ocean-loving cousins, live under stones, in empty shells, and in tubes they build in the sea.

Members of the roundworm family live in ponds, deserts, and even ice.

Many kinds of worms are **parasites** (PEAR uh sites). Parasites live in or on a living animal and injure or kill it.

This colorful bearded fireworm lives in the sea

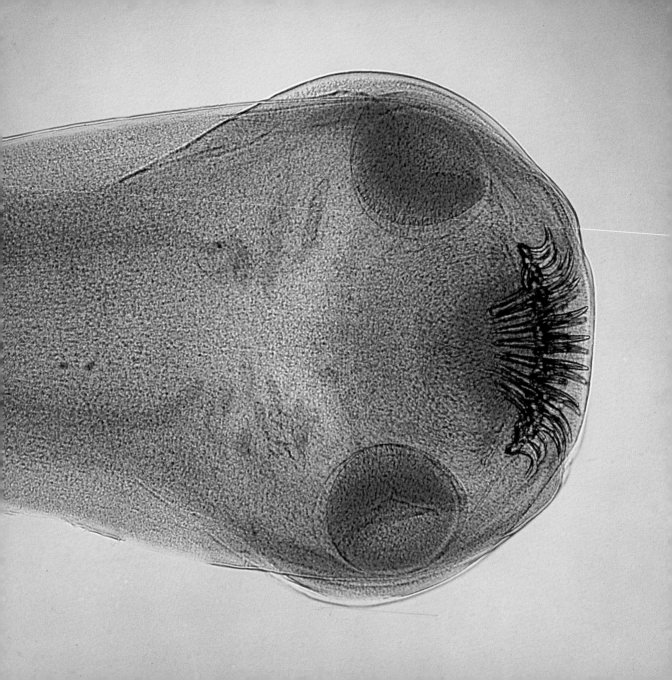

WHAT WORMS EAT

Parasite worms more or less feed on the animal—or person—that is stuck with them. The parasite's victim is called the **host** (HOST). The parasite usually lives on its host's flesh or blood.

Earthworms and many other worms are not parasites. Earthworms eat tiny bits of plant or animal matter in the soil. They swallow soil and filter out the goodies.

Certain marine worms have a long, sharp snout that they use to kill the sea creatures they eat.

This is the head and mouth of a tapeworm—magnified 40 times—that lived in a cow

A California state park's aid lifts a marine worm

Not ready for lunch, a toad watches a nightcrawler slither between its legs

KINDS OF WORMS

Scientists are still finding "new" worm species, thousands of them. Scientists have already identified over 15,000 species of roundworms, but there may be another half million species of roundworms!

The earthworm family has at least 1,800 members. The common nightcrawler of North America is a medium-sized earthworm. The largest is a nine-foot long Australian earthworm, a worm large enough to scare a fish instead of catch it!

These hookworms were parasites of a dog

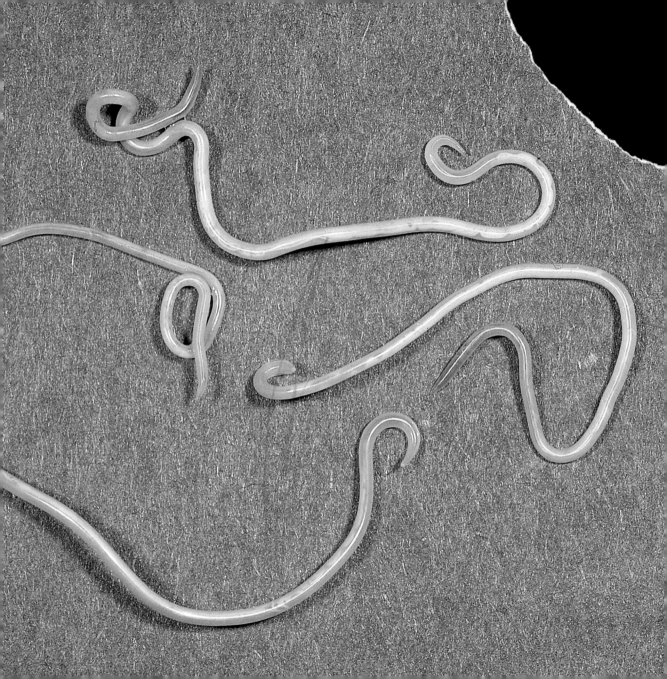

EARTHWORMS

Earthworms are the best-known worms in North America. They crawl onto the surface of the soil at night in rainy weather. Sunshine and dry weather drive them back into their burrows.

Earthworms escape cold and winter by burrowing deep beneath frozen soil. There they gather in large, wriggling, slimy masses.

Worm slime is a liquid called **mucus** (MU kuhss). One reason worms make mucus is that it eases a worm's crawl through its burrow.

Soft and squeezable, earthworms can be stretched, too, but they will snap if stretched too far

SOIL BUILDERS

Earthworms are helpful to human interests far beyond being fish bait.

Earthworms are important in keeping soil healthy for plant growth. Earthworm tunnels make it easier for plant roots to spread into the ground. The tunnels also help water enter the soil.

By swallowing soil, earthworms help mix soil ingredients. That, in turn, makes the plant foods in soil easier for growing plants to use.

By burrowing into and "eating" soil, earthworms are important soil builders

EARTHWORM ENEMIES

Every boy or girl with a fishing pole may be an earthworm's enemy. Every farmer who spreads fertilizers and insect poisons into the ground is a much greater enemy. Chemicals in fertilizers and poisons kill earthworms.

On American lawns, the robin is the enemy. Robins love to yank fat earthworms from the soil. Snakes, centipedes, salamanders, and large beetles also snack on earthworms.

A garter snake finds an earthworm lip-smacking good

PEOPLE AND WORMS

Earthworms are not only harmless to people, they are helpful. Certain kinds of parasite worms, however, are a major problem. Worms called blood flukes, for example, affect over 200 million people in parts of Africa and South America.

Other kinds of worm parasites also attack people along with pets and wild animals. The tapeworm family is made up entirely of parasites.

Glossary

host (HOST) — a plant or animal on or in which a parasite lives

invertebrates (in VERT uh brayts) — the simple, boneless animals such as worms, snails, starfish, and slugs

marine (muh REEN) — of or relating to the sea

mucus (MU kuhss) — a thick, sticky liquid produced by worms and other animals for various reasons

parasite (PEAR uh site) — an animal that lives on or in another living thing, feeds upon it, and finally injures or kills it

species (SPEE sheez) — within a group of closely related animals, one certain kind, such as a *red* fox

INDEX